TIME IS SLIPPERY

INTRODUCTION

I was lost in my dreams
a long time ago,
and
never came back.

Time is certainly a slippery concept; for example, even the direction of time, the division of time into past, now, and future, called the arrow of time by physicists is slippery. Most physical laws are written in terms of differential equations which are mathematically reversible with respect to time so that there is no real distinction between past and future. One of the few exceptions is the Second Law of Thermodynamics which states that the entropy of a closed physical system tends to increase with the passing of time.

The Second Law was originally framed in terms of heat transfer and work; and appears to be designed to preclude the possibility of a perpetual motion machine. Historically, the Second Law first appeared in the context of explanations of steam engines and their efficiencies in the first part of the nineteenth century. Entropy was defined mathematically in terms of heat transfer and temperatures, there was no thought of making any deep connection between entropy and time, and the modern idea that the arrow of time is defined by the direction of increasing entropy would've seemed nonsensical. With the connections between thermodynamics and dynamics made by Boltzman and Gibbs in the late nineteenth century entropy appears to be far better understood than time. The point of this essay is to describe in some detail just how hard it is to define time and how slippery the idea of time is. Entropy, on the other hand, is fairly simple if one looks closely at the gas or steam trapped inside the cylinder of a steam engine. In this case the physical system to be described is not only the steam inside the cylinder, but also the pressure exerted on the sides of the cylinder, the temperature of the gas inside the cylinder, or any other macroscopic state of the gas-cylinder system. The gas inside the cylinder is made of millions of small particles, each of which has a particular position and momentum according to classical dynamics. The dynamics of the gas is given by writing down the positions and momenta of each particle; and a microstate of the gas system is given by a particular combination of all the positions and momenta. A probability is assigned to each configuration; and entropy is a measure of these probabilities, low entropy corresponding to low probabilities and high entropy, to high probabilities. For example, a situation in which all the particles are lined up against one side of the cylinder and are proceeding in one direction is a situation with very low probability and entropy. According to Boltzmann:

"Entropy is a measure of the number of particular microscopic arrangements of atoms that appear indistinguishable from a macroscopic perspective."

This definition isn't entirely clear though it seems clear enough as far as arrangements of atoms are concerned. At the time Boltzmann wrote this definition it had to be fairly daring because atoms were a conjecture somewhat like black holes or strings today, yet now it appears 'arrangement of atoms' is one of the few solid parts of his definition. The words 'microscopic" and 'macroscopic' aren't clear today because of the obfuscations and muddle at the center of quantum theory, the reigning king of theoretical physics. According to modern theory microscopic refers to physical regions so small quantum theory applies, while macroscopic refers to regions so large that either classical dynamics or general relativity applies. The connection between microscopic and macroscopic has its own theory, measurement theory. Measurement theory is a grand muddle ranging from the solipsism of demanding the existence of a conscious observer, whatever that means, and the extreme realism of Everett's multiple worlds or universes.

Time is intimately related to entropy because increasing entropy is taken to define the direction of time, time's arrow as in the Rubiyat of Omar Khayyam:

"Time flies like an arrow, fruit flies like a banana."

Increasing entropy defines the relative positions of past, now, and the future for the kinds of minds that take theoretical physics seriously. In the past entropy was relatively lower than now, and in the future entropy will be even higher.

The usual explanation of entropy could be called the Dropped Egg Analogy. Imagine dropping an egg on the floor. In the beginning, as the egg leaves the egg droppers hand the entropy of the egg physical system is low; and finally the entropy of the smashed egg system on the floor is high. In other words, entropy is a measure of order and disorder which applies to egg systems, and if it applies to egg systems it must apply to steam engines, black holes, the universe, and even whole clutches of universes. Since the egg system never gathers itself off the floor and flies whole up to the egg dropper's hand, the march of entropy corresponds to the march of time although only in the sense that the march of entropy defines the direction of time. The trouble with the Dropped Egg Analogy is that order and disorder are never defined mathematically or precisely in the same manner that macroscopic and microscopic are never defined precisely in Boltzmann's definition of entropy. And what does a dropped egg have to do with steam engines, black holes, etc?

The idea of entropy is extended far beyond a dropped egg or a steam engine and applied to both black holes and the entire universe among other things. In the case of black holes or universes it's not entirely clear what corresponds to or generates microstates and macrostates, or order and disorder. However the mathematics is extended to apply to what might be imaginary objects such as black holes or multiple universes without anyone ever having seen either one. Of course no one has ever seen an atom with the naked eye either; but the evidence for their physical existence seems to be very good. The power of mathematics to describe known physical objects and systems encourages the assumption mathematics can describe systems such as black holes or multiple universes as long as the mathematics is done correctly. Perhaps this is a trap. Physical reality isn't necessarily the same as mathematical reality.

The increase of entropy only gives a direction to time and does not answer many other questions about time. What is the proper way to measure time? Does the nature of time change from one physical system to another? Does time even exist?

In the case of the steam engine piston system equilibrium is reached very quickly if the piston is allowed to remain stationary; in other words, an equilibrium value of entropy is reached very quickly. Why isn't the entire physical world already at equilibrium, and why should entropy be increasing? Why should the arrow of time point anywhere at all? These questions are avoided by invoking the Big Bang, a low entropy situation at the very beginning of time, and a finite amount of elapsed time, 14 billion years or so, since the Big Bang. However, this is not a deep answer because it brings up the question of why entropy should be low at the beginning; or, even worse, why there should be a Big Bang at all.

As far as the rate at which time enfolds, or the choice of a good clock, it seems to depend on the overall system involved. For example, what is the best way to time the enfolding of events on the screen of an individual computer; or, what is time as far as the individual computer screen is concerned, individual computer time? Apparently individual computer time depends on a computer's internal clock, the clocks of its various servers, the transmission times on the computer's network, the particular path traversed from the momentary server to the individual computer, etc.

Many years ago I was keeper of the crackpot file at a university which will remain anonymous. I attained this high position due to natural talent rather than any professional achievements. Many of the items in our crackpot file were priceless. For example, one gentleman submitted a lengthy paper proving geometrically that Newton was wrong, that force is really proportional to the inverse cube of the acceleration of a body. On the back of this lengthy paper he had written a very good modern version of Don Quixote.

Most of the crackpot contributors appeared to be very interested in physical ideas and physics which is more than could be said of some of the graduate students and professors who were simply holding a job, a place in the system. A few cranks have appeared to counter the current orthodoxy in physics, the current dead-end. There are some professors who are interested in the fundamentals of quantum theory and other esoteric subjects such as the philosophical and mathematical study of time. As a matter of fact, there are a lot more professional questioners of the nature of time and quantum theory now than there were when I was in charge of the crackpot files nearly fifty years ago. In the early sixties many of the big-shots in physics departments were brutal braggarts who gloried in their sojourn at Los Alamos during the construction of the first nuclear weapons. Very few braggarts ever question anything.

Many people have commented on the peculiar ability of mathematical constructs and ideas to describe physical reality. Here the idea is advanced that time is the key mathematical construct that allows a description of physical reality. Unfortunately, time is slippery.

Does Time Exist?

Quantum fluctuations in the Sea of the Vacuum,
the Sea of Violent Nothing
on which everything sails like a ghost ship
made of dreams of dead pirates.

Perhaps all that can be said is that time is a way to measure change, what kind of time to use as a measure depends on the kinds of change and the physical processes involved. For example, in discussing local galactic dynamics it might work to Use the spinning Milky Way as the proper clock. Although the difference between this kind of time and our local lab time is usually thought to be caused by dark matter and energy, the true cause of this difference may be more subtle and hidden from us by our ignorance by our ignorance of the proper concept of time and its relation to physical change and processes. We really don't understand time.

Time can be taken to be a philosophical, metaphysical, or purely mathematical object in the same sense the integers, rational, and irrational numbers are mathematical objects. If time is taken to be a mathematical object, Kurt Godel has an interesting argument showing time does not exist. His argument is the often used reductio ad absurdum method of proof used by mathematicians. This form of argument assumes the opposite of the thing to be proved and shows that this leads to an absurdity, something that can't be true. The most famous example of this mode of argument is Pythagoras' proof that the square root of 2 can't be a rational number; that is, a number which can be written as one integer or whole number divided by another whole number such as ½. Rather than accept the existence of numbers beyond rational numbers, Pythagoras and his followers hid this nasty fact, which perhaps entitles them to be thought of as the first modern thinkers or intellectuals. With Pythagoras, assume the square root two can be written as a ratio of two whole numbers and square the result:

$$a^2 = 2b^2 .$$

In addition suppose all common factors have been cancelled out, something that can always be done according to elementary arithmetic. Two times any number is always an even number, therefore a squared in the above equation is also an even number. But if a squared is an even number then a itself must be an even number

$$a = 2c.$$

Squaring both sides gives

$$a^2 = 4c^2$$

or

$$2b^2 = 4c^2$$

or

$$b^2 = 2c^2.$$

But if b squared is an even number, as this last equation states, then b must be an even number. Therefore both a and b are even numbers which is not possible because we cancelled out all the common factors between a and b. Since it is always possible to cancel out common factors in a ratio, it must be impossible to express the square root of two as a ratio of two whole numbers.

Godel's proof of the nonexistence of time follows the same kind of pattern as far as the logical argument goes, he uses a reductio ad absurdum argument which first assumes the existence of time and then shows it leads to untenable or absurd results. This argument is more compact than Pythagoras' proof of the irrationality of the square root of 2. Assuming time exists, look at the solutions of Einstein's differential equation covering general relativity. It is possible to find a form of the stress-energy tensor on the right-hand side of the equation which yields closed time-like curves. In other words, it is possible to have a physical situation in which one is trapped, doomed to repeat the same history continually, as in the popular movie Groundhog Day. This is absurd, and since Einstein must be right, time must not exist.

But is Einstein right? There does appear to be one spot in his theory of special and general relativity which attracts an inquiring, skeptical mind. This blemish, if it is a blemish, certainly appears to be very minor compared to the disaster called measurement theory in quantum mechanics. Relativity uses light as a synchronizing signal, which determines the nature of space and time. Why not use some other kind of signal such as a sound wave which might be used by a dolphin scientist since most of his world is composed of images he generates with his sonar? Although this might seem absurd, it certainly isn't a logical absurdity. Should the synchronizing signal fit the physical situation in some unknown manner?

Although nearly all of the information we have about astronomy comes by way of some part of the electromagnetic spectrum, can we be sure light behaves in the far reaches of the cosmos as it does here? What if the synchronizing signal is changing with time? Doesn't the idea of simultaneity and the present become even more nebulous?

Kurt Godel wasn't interested in criticizing his very good friend, Einstein. His focus was on time itself and he assumed Einstein's theory was correct. His conclusion was that time as we understand it does not exist. I have no idea whether time exists or not; questions seem far safer to me than conclusions or answers.

Kurt Godel

Time stretched,
changed into the time of wandering forms,
of mice turning into men,
of wandering continents and sheets of ice.

Kurt Godel was a German speaking refugee from Nazi Europe, a Viennese Lutheran with a conscience, who found himself stranded in New Jersey at the same institution as Albert Einstein. They became fast friends although they composed an odd couple. Godel was a hypochondriac whose malady contributed to his death, while Einstein kept his physical condition secret because he didn't feel it was worth screwing about with it. But both were revolutionaries; Einstein in physics, Godel in mathematics.

Godel was not known for his work covering the nature of time, but for his theorem on the completeness of axiomatic systems. This theorem destroyed much of Bertrand Russell's Principia Mathematics in which Russell spent much labor proving that one plus one is two from axiomatic principles of mathematical logic.

Imaginary Numbers

Time changed into the cycles of the universe,
 cycling through the Sea of Violent Nothing
 into different universes
 with electrons as big as elephants,
 men made of galaxies,
 crystal squid swimming in mercury seas.

Though it seems to be fairly easy to ask decent questions about time as far as relativity is concerned, asking decent questions about quantum time is a different matter. First of all, what's the mathematical structure of quantum theory? The central concept is the superposition principle, which refers to the linearity of the wave equation. Any linear combination of two solutions is also a valid mathematical and physical solution. In general solutions of the wave equation are complex or imaginary functions of a real variable. A solution of the wave function gives a probability amplitude which must be squared to find a physical probability function which contains the physical information.

Probability theory can be slippery enough without having to deal with an imaginary probability amplitude. Several years ago a quiz program posed a probability problem which even stumped a lot of professional mathematicians. Imagine three closed doors. Behind one of the three is a crock of gold, behind the other two doors is a crock of shit. In your mind you pick one of the doors, but before you declare your choice one of the doors is opened to reveal a crock of shit. Would your changes be improved, assuming a random distribution of gold and shit, by changing your choice? Some mathematicians answered No since if the distribution of shit and gold is random no previous trial gives any information about the next trial. For example on a roulette wheel the chances of a red or black on the next turn should still be even if black had turned up ten times consecutively on previous turns. The correct answer to the crock problem is Yes since the probability of picking the right door before the first crock is revealed is one out of three, but the probability of picking the right door after the revelation of the crock of shit is one out of two. In the crock case, a previous trial does yield valuable information. Adding imaginary or complex number theory to this slippery stuff just makes things worse; and the probability amplitude which is the key of quantum theory is a complex function of real variables describing position and time.

My achievements running the physics crackpot file at the large state university may have been the reason I was selected to teach a course in imaginary numbers at a small college in Kansas City. Most of my acquaintances thought I was getting away with an immense scam, that I was being paid to teach a course in something that didn't exist, that was imaginary. Of course some of these acquaintances also thought I had wasted my life entirely by studying quantum theory, that I would have been much better off studying for ten to twenty in the local penitentiary, one of the best finishing schools in the country. Evidently our political elites agree with them since they never encourage the study of quantum theory to the extent they encourage attendance at one of our fine penitentiaries for ten to twenty. At the time I entered college Sputnik and the reverence accorded physical scientists for their creation of nuclear weapons had put a gloss and glow on the study of physics which I was not able to resist. During the late fifties even the backwater physics departments in which I splashed were quite large. Fifty years later those same departments are now sad, tiny remnants if they exist at all since they siphon funds away from really important programs in business, marketing, and finance while the main institutions of the corporate police state, the prisons, have become giants. The U.S. has 25% of the world's prison population with only 5% of the world's total population. Yet at the same time fundamental physical science is neglected, our elites seem to believe technology is capable of anything, that technology is capable of overcoming any of the problems of resource depletion, that technology has given us an Earth with infinite resources even though Earth itself appears to be nothing more than a small blue dot swimming in infinite black space.

In the past fifty years educational standards in the U.S. appear to have fallen to a new low. The neglect of physical science is only a small part of over-arching neglect and corruption, of the conversion of the school system at all levels to a system of interlocking corporations dedicated to power and money, which of course is far more important than learning or scholarship. In the Kansas City area it's common for high school graduates to read at a second-grade level. No doubt graduates from the local colleges can read at the third-grade level, which means this little essay on time is inaccessible to them since it is written on a fourth-grade level. On top of all this the poor student is forced into irremediable debt to pay for his defective education and can not find relief in bankruptcy court. He must pay no matter how bad his education may be.

Actually there is nothing mysterious about the use of complex variables in quantum theory, it is even common to use complex variables in electrical engineering. The most general solution of the wave equation can be expressed in complex variables. In order to satisfy the superposition principle of quantum theory, the algebra of imaginary numbers must be used to combine two solutions of the wave equation. The superposition principle is more fundamental to quantum theory than the form of the wave equation. The form of the wave equation depends entirely on the physical system which is being described by quantum theory, and in some cases it's hard to figure out exactly what the physical system is. In the case of cosmology, the physical system is the entire universe, everything that exists, has existed, or will exist. But what exactly do these words mean as far as the mathematics is concerned. In the same sense, what is the vacuum, that which is exactly nothing. How is nothing described mathematically? How does anyone measure the properties of everything or of nothing?

Newtonian Time

Time dropped to the ground
and splintered into myriads
of different times,
of different seasons.

Newtonian time is absolute time, a kind of time with
no attached nonsense. Newtonian time is absolute in the
sense that now and the past and future are not in question.
Although "now" may be 12:00 P.M. in Kansas City and 8:00
A.M. in Honolulu, the different times only refer to
different time zones. Now is the same time in Kansas City
and Honolulu no matter what words are used to describe now.
By contrast, now in special relativity is not the same
everywhere in space. Strictly speaking there is no
universal now in relativity theory.

Newtonian time is also ordinary clock time which
proceeds tick by tick from now till the end of time in
the future, or which proceeds tick by tick forever.
Absolute time does not speed up or slow down but
proceeds regularly forever. In addition, absolute
time can be divided into intervals of any size; there
is no shortest or longest time interval. This kind of

time allows definitions of both velocity and acceleration as time derivatives of position in space; and, therefore the development of Newtonian dynamics and astronomy.

Newtonian time is also used in quantum theory. Most of the applied results in quantum chemistry and physics use Newtonian time. On the other hand, relativistic quantum field theory can only use absolute time as an approximation. There is even some question, as we shall see, whether or not it's appropriate to use absolute time in the case of ordinary non-relativistic quantum theory.

In 1930 in Brussels at the sixth Solvay conference Einstein and Bohr, the strongest apologist for the new field of quantum theory, had an important argument over the nature of time in the new theory. The argument involved the nature of the time-energy uncertainty relations in quantum theory. According to Bohr's interpretation of quantum theory the results of measurements of physical quantities are numbers that occur in differential equations, and the physical quantities themselves such as position, momentum, time , and energy are represented by differential operators. This implies an uncertainty relationship between energy and time, in particular it relates line widths for unstable physical systems (radiating atoms) with the corresponding half-life times. Einstein did not believe in quantum theory and came up with the following counter example, the following gedanken experiment. Consider a box filled with radiation. A clock controls the opening and closing of a shutter at a particular time so only one photon escapes the box. If the box is weighed before and after the escape of the photon, both the energy and the time of release of the photon are known, contradicting quantum theory's uncertainty relationship. Bohr's rebuttal used Einstein's redshift formula from relativity to restore the uncertainty relation. It appeared Einstein had been hoist upon his own petard. Afterwards Einstein ceased trying to find inconsistencies in quantum theory and focused on its possible incompleteness. Modern believers in quantum theory believe it to be complete in the sense it correctly describes atomic systems, billiard balls, the entire universe, and absolute nothing or the vacuum. This requires quite a stretch of the imagination as will be seen.

Relativistic Time

Then you die and time stops,
doesn't seem possible
does it?

In some ways time is like space. History books and calendars define a geometrically ordered time series with definite dates and times just as a ruler has a geometrically ordered series of points in space labeled as one centimeter, one meter, etc. However, a clock measures "now", an entirely different property than a point in space or a point in time in the future or the past. "Now" is commonly the origin of both past and future, the origin in any plot of points of time. From an intuitive point of view there must be a universal now, a universal, worldwide present or now. Right now something is happening here, in Moscow, on Mars, at the center of the Milky Way. But how is "now" determined at different spots in space? In order to determine now at different points in space there must be some means of synchronizing clocks at different points, some signal which can be used to synchronize clocks. In special relativity the signal is light.

Therefore the motion of light through space is central to the construction of time throughout space. And unfortunately, the motion of light through space is such that it isn't possible to construct a worldwide now.

The time that appears mathematically in special relativity incorporates the manner in which light signals move through empty space. If a different signal than light were used to synchronize clocks, a signal with infinite speed or Galilean properties, a different kind of time would appear. A signal with infinite speed would bring us back to absolute Newtonian time, a time with a worldwide now. The reason a light signal is used is that it is one of the few signals that traverses empty space although it is not the only signal that can traverse empty space. High speed bursts of particles from any of stars also traverse empty space. The idea is to find the best signal to synchronize clocks, the signal which best incorporates the fundamental essence of both relativity and quantum theory. In the case of special relativity alone the light signal seems to be the correct synchronizing signal.

What is particularly interesting about Einstein's relativistic time is that it led him to the construction of an entirely different idea of both space and time. Time appears to be the key element in the space-time of special relativity.

Originally the time appearing in quantum theory, the

time variable in Schroedinger's equation, for example, was absolute time. Later Dirac developed a wave equation which was relativistically invariant and which appeared to use the time of special relativity. However, the strict application of the uncertainty principle of quantum theory to the fabric of space–time implies there is no "now" or even "before" or "after" on a small scale. This situation is even more radical than the lack of a universal "now" that appears in special relativity. The use of light as a synchronizing signal fouls up any intuitive idea of time, and the application of quantum theory finishes our simple ideas of time completely. Unfortunately there is no good mathematical definition of time either. The time that works for special and general relativity doesn't work for quantum theory. The disconnect between relativity and quantum theory has been around for nearly a century now. The usual statement is that quantum theory governs the physics of the very small while relativity governs larger objects such as stars, planets, and galaxies; but there are few connections between the two realms of size other than some connections established by statistical mechanics which relates myriad microstates to average macrostates.

Most believers in modern quantum theory insist that quantum theory is valid everywhere, for all times, and for all sizes. That is, they believe quantum theory is a complete theory of physical law. For very large astronomical sizes quantum theory is also valid, but is restricted to a form which reduces to classical dynamics or general relativity; and that it is more appropriate to use classical dynamics than quantum theory because of the nature of the physical systems involved. Modern theorists are forced into this position because they certainly wouldn't want to agree with the wrong-headed Einstein that their theory is incomplete. They must also argue that the vacuum, the state of nothing, is correctly described by quantum theory. But conventional quantum theory involves the study of one particle, or of several particles interacting with each other. The number of particles in any quantum mechanical system can vary; but what is really distressing is that the vacuum state of any system is not well defined. The particle concept does not have universal significance since particles may register on some detectors, but not on others depending on how the particle detector is moving. The acceleration of a particle detector can cause the appearance of particles. Also, the presence of virtual processes within the quantum mechanical vacuum complicate the matter even further and cause renormalization effects in the properties of single physical particles. The extent to which quantum theory must be stretched to account for the state of no particles is hard to believe. This kind of imaginative stretch is not the hallmark of a complete physical theory.

Cosmology

I was tired of lying there dead,
I got up and walked off,
into the land without clocks.

As a complete physical theory, quantum theory should include a theory describing the entire universe. Therefore it should be possible to write down a wave function, a probability amplitude for the entire cosmos. It is easy to write down a bunch of symbols and call it a wave function; but what does it mean? How is it connected to physical reality? In particular, what does a measurement of the cosmic wave function mean?

In conventional quantum theory a measurement causes the wave function to collapse into a state corresponding to a particular physical state. The prototype of this collapse is used to explain the nature of diffraction experiments involving either photons or electrons, and to connect the microscopic realm of the wave function with our macroscopic realm. A diffraction pattern is formed when waves or photons from two different sources interfere. The pattern is formed on a photosensitive screen one point at a time even though the pattern covers a relatively wide area. If the intensity of the photo beam is reduced until only one photon can be between the two sources and the screen at any one time the diffraction pattern's form is not altered. It appears the photon is interfering with itself like a wave form until it reaches the photosensitive screen where it collapses to a single point, ejecting a single electron from a particular atom in the screen. The wave function describing the photon is taken to describe a single photon to accommodate the photon's apparent interference with itself. According to this interpretation, the Copenhagen interpretation after Niels Bohr, the wave function of the photon collapses to the kind of wave function associated with a single position when the photon hits the screen. When any measurement is made the wave function collapses to the kind of wave function associated with that measurement. Many students find this explanation of photon and electron diffraction and of wave function collapse hard to believe. It's hard to believe an electron, whatever an electron really is, can shift from wave like existence to particle like existence at the drop of a screen or other detector.

Since quantum theory is a complete physical theory, it must be asked what happens when the wave function of the cosmos is subjected to a measurement. Does the wave function collapse to a single kind of universe among a whole multitude of possibilities? How can a wave function describing everything be measured if any measuring apparatus is to be outside the physical system being measured, and the system being measured includes everything?

Hugh Everett avoids these problems by avoiding collapse altogether. Everett assumes separate perception of coexisting but dynamically decoupled branches of the wave function. This is described as branching of the consciousness of the observer. Each branch of the wave function describes an alternate version of reality, an alternate universe. Unfortunately the branching of the consciousness of the observer is as mysterious as the collapse of the wave function on observation. But Everett's quantum theory fits the requirements of quantum cosmology. Since no observer can be outside the universe, any quantum mechanical observation within the universe simply branches the universe. Everett's quantum theory suggests a super universe, the union of all possible universes. Philip K. Dick expanded on the idea of multiple realities in much of his science fiction, piling one paranoid reality on top of another in dreadful profusion.

Although Everett's multiple realities is a minority interpretation of quantum theory it is still taken seriously. On the other hand, Einstein's contention that quantum theory is an incomplete theory requiring baffling conceptual extensions to deal with many parts of physical reality is not taken seriously at all.

Measurement Theory

As I ran into the past
the houses disappeared
my arms became matted with hair
and dragged on the ground.

According to standard measurement theory, a measurement
of a microscopic system such as an electron or proton
causes the wave function of the microscopic system to
collapse to a wave function that is peculiar to that
measurement. A measurement on a microscopic entity
requires a much larger measuring device, a macroscopic
apparatus of some kind which can be read by an observer.
Since everything should have a wave function if quantum
theory is a good theory describing everything rather
than a rule-of-thumb for extremely small objects, the
macroscopic measuring device should have a wave function
also. Therefore the act of measurement causes both the
wave functions of the measured and measuring systems to
collapse to a wave function corresponding to the
measurement observed. If nothing is said about the
physics of the collapse, if nothing is known about this
collapse, the only information available is the observed
value. In other words if there is no one to make an
observation nothing is known about either the small or

large systems. In particular, it's impossible to say whether the systems collapse or not. To avoid this kind of solipsism and provide a definite mechanism for connecting small systems with much larger systems, a different, stochastic evolution of the wave function is sometimes used. Collapse in these systems occurs automatically without the aid of an observer whenever a microscopic system is entangled with a macroscopic environment as is typical for a measurement. Since the collapse is a definite, well-defined process it can be observed also and thus leads to deviations from the usual predictions of quantum theory.

Aside from the problems of size involving what is small and what is large, there are problems defining the systems themselves. An electron polarizes the surrounding matter or vacuum, the surrounding environment, leading to an effective "dressed" particle with a renormalized charge and mass equal to the measured charge and mass of the electron. A clock based on the radiation from such a dressed orbital electron rests on the properties of both the electron and its environment.

A particle's environment includes its space-time metric, its vacuum. In quantum theory the vacuum metric is taken to be flat and time is taken to be Newtonian time. Unfortunately the fourth diagonal element of the metric, the pure time element, includes a factor of the square of the speed of light. The square of the speed of light is an enormous number; therefore any small fluctuations of the time part of the metric could be large also. This effect is generally neglected by using a system of units in quantum theory in which the speed of light is set equal to one. The square of one is still one.

The root of the problems of measurement theory seems to rest in the problems of going from a small to a large system. In a theory like thermodynamics all observed quantities are averages, but in quantum theory the properties of small systems can be read directly in our large instruments. This strange situation has led to some very strange proposals such as Everett's formulation of quantum theory.

Decoherence

I launched the kite
up into the night wind,
when I pulled the string
in out of the darkness
there was nothing at the other end.

One of the strangest conceptual problems has to do with the classical limit of quantum theory. Each new and improved physical theory should include the old theories in some kind of limiting case. For example, special relativity approaches Newtonian theory for small particle velocities. Special relativity has a definite "classical" limit in which the dynamics of relativity approaches classical dynamics. Quantum theory does not have a definite classical limit in the sense that an increase of size or mass leads to macrophysics or classical dynamics in an obvious manner. In quantum theory increasing the mass of a body does not rule out coherence of the wave function; that is, the body in question behaving as a wave rather than a point particle. Entire books have been written attempting to create a classical limit for quantum mechanics. Springer Verlag published a book about a decade ago with the title "Decoherence and the Appearance of a Classical World in Quantum Theory." Decoherence is an imaginary technical term which begs

the question of a classical limit in that decoherence is
is the process whereby the environment becomes entangled
with the physical system and any measurement devices in
such a way to produce the appearance of a classical
result. Decoherence is a process reminiscent of the
addition of epicycles to correct the painful discrepancies
of Ptolemaic astronomy of several thousand years ago.
In the case of the entire universe, quantum cosmology,
there is no environment. The physical system in this case
is the entire universe and there can be no environment.
In the case of a comet traveling through interstellar
space, the environment is the background radiation from
the Big Bang. If not for the primordial Big Bang, the
comet's wave function could easily cohere and the comet
would become a large quantum object capable of diffraction,
tunneling, and all the other eccentricities of ordinary
quantum objects.

However even classical objects are not what they seem.
Chairs and tables are classical objects which follow
classical trajectories if thrown, but chairs and tables
are mostly empty space and only their quantum properties
allow them to appear to us as solid objects. Quantum
theory was postulated to explain the stability of matter
consisting of large groups of atoms made up of electrons
circling nuclei that together make up classical objects
such as chairs and tables.

Therefore it takes large conceptual stretches to retrieve the vacuum, astronomical objects, and classical objects such as chairs and tables out of quantum theory. Yet quantum theory is even stretched to include gravity, to include what is called quantum gravity. Perhaps the admission that quantum theory is not a complete theory would argue against trying to stretch it to include gravity; but modern theorists can not admit that Einstein was right about quantum theory.

The Speed of Light Equals One

"There you see that portion of the lunatics who are not working on the quantum theory."

A. Einstein to P. Frank outside the nuthatch in Prague, 1912.

Actually in terms of miles per hour or kilometers per second the speed of light is a very large number; but particle physicists, with their usual disdain for big numbers, pick a unit system in which the speed of light equals one to simplify their quantum mechanical expressions. Particle physicists also don't seem to care that their accelerators have very big numbers on their price tags. However, if there is a very small variation in the fourth component of the metric tensor the variation will be multiplied by this very large number at least once. The fourth diagonal element of the metric tensor contains a factor of the square of the speed of light, an extremely large number. If variations in the fourth diagonal element of the metric tensor, as suggested by the superposition principle, are used to set up an alternative to ordinary quantum mechanics, the wave function for a particle, or what appears to be a particle, becomes a sum over different spaces characterized by different values of the fourth diagonal component rather than a sum over different eigenfunctions defined on the same flat space. Therefore the particle structure is intimately connected with the structure of its surrounding space-time, a global mathematical feature. From this point of view effects such as the Aharonov-Bohm or EPR phenomena become more comprehensible than they are from the point of view of ordinary quantum theory where dispersions or variations in values are restricted to a matter field.

Quantum Gravity

"Quantum theory is all bullshit."

Larry Mackey, Dave's Stagecoach Inn, 2002

The conceptual difficulties of quantum theory compared with the lack of conceptual difficulties in relativity theory suggests quantum theory should be changed to incorporate relativity rather than vice-versa. For example, a hazy space-time with a family of wave functions corresponding to a family of different metrics could be postulated. The crucial variation in the different metrics would be in the parts of the metric associated with time because of the enormous size of the speed of light. In turn the different metrics define a family of times corresponding to each possible metric, a multifoliate time. Evidently the amplitudes associated with each time can be picked so that quantum mechanics is approximated. However, instead of a wave function formed from a sum of eigenfunctions defined on the same flat space-time, a hazy wave function would include functions defined over different spaces characterized by different values of the metric. Therefore the particle structure is intimately connected with the structure of its surrounding space-time, a global mathematical feature. A measurement would pick out one of the times included in multifoliate time instead of collapsing the wave function to the correct eigenfunction corresponding to the measurement or by a branching of reality as in Everett's quantum theory.

In other words quantum gravity may require multifoliate time or some other kind of time. Classical physics uses absolute, Newtonian time, relativity uses the time of Einstein. Quantum gravity may require a different kind of time. There seems to be no point in philosophical arguments about whether or not time exists. Time is an abstract construction that has proved essential to the description of physical reality. But what kind of time fits what kind of physical reality?

Quantum Gravity and Cosmology

"Doing time in prison is better than doing time in a college because you learn more and aren't left with a huge, unpayable debt that must be paid."

Larry Mackey, Dave's Stagecoach Inn 2002

What is nothing, or to put it slightly more technical terms, what is the vacuum? According to ordinary quantum theory the vacuum seethes with quantum fluctuations of various kinds including self-destructing pairs of particles and antiparticles. According to conventional cosmology the big bang was the result of a quantum fluctuation that burst from nothing into the entire universe. At the precise moment of the big bang not only mass and energy were created; but also time and space themselves. Before the big bang, even though it is impossible to talk about a time before the big bang since time didn't exist before the big bang, nothing didn't exist either since space itself did not exist according to the conventional theory. It's certainly good that it makes no sense to talk about a time before time existed.

What sense does it make to talk about a quantum event, the big bang, which supposedly created space and time when we have no coherent theory of quantum gravity, no coherent theory of creations or destructions of space and time? What sense does it make to talk about the wave function of the universe when it is not known how to include space and time? Therefore what sense does it make to add dark energy and mass to cosmological theories in an ad hoc manner reminiscent of Ptolemaic epicycles? Perhaps a good quantum gravity would decohere into several classical or semi-classical limits corresponding to different light velocities and spaces which would explain the bizarre movements of galaxies and star clusters without assuming dark energy or mass.

Perhaps we need to be much more careful about fitting the mathematical framework for time to the actual physical system to be described. For example, Einstein's criticism of the use of a Weyl space to unify gravity and electromagnetism involves the movement of a large clock, a macroscopic clock, and perhaps Weyl spaces apply only on a very small scale like quantum theory.

At any rate, we need more understanding than afforded by quantum theory, a grotesque, hundred year old rule of thumb which does more to obscure than clarify.

Does Time Exist?

Whether time actually exists or not is a philosophical or even metaphysical question. Einstein's relativistic time is not the same as Newton's absolute time. Perhaps all that can be said is that time is a way to measure change, what kind of time to use as a measure of change depends on the kinds of change and physical processes involved. For example, in discussing local galactic dynamics it might work to use the spinning Milky Way as the proper clock. Although the difference between this kind of time and our local lab time is usually thought to be caused by dark matter and energy, the true cause of this difference may be more subtle and hidden from us by our ignorance of the proper concept of time and its relation to change and physical process. Similarly there are difficulties constructing ordinary lab time from the motion of a large body composed of many small particles if multi-foliate quantum theory is used. Time is certainly very slippery.

Hidden Variables

"American universities have priced themselves out of the market. How many will survive?"

Anon.

One of the peculiarities of quantum theory is that it describes a group of measurements performed on a group of physical systems; it does not describe a particular experiment performed on a particular physical system. As a matter of fact, quantum theory says nothing about one measurement performed on one atom or whatever. Niels Bohr, one of the creators of quantum theory, argues that it is impossible to know anything about one measurement on any one quantum system; it is only possible to make statistical statements about many measurements on many similarly prepared systems. In other words, the solution to the absurdity of Schroedinger's cat problem is there is no problem according to Bohr's interpretation of reality. It makes no sense to talk about a single, particular cat that is half dead and half alive at the same time; it is only possible to talk about the statistics of a herd of cats.

As Lee Smolin points out in "Time Reborn", a hidden variable interpretation allows a description of a single, particular event by assuming a guiding wave for particles, which propagates instantly, thus violating relativity. In a hidden variable theory there is a preferred notion of no movement, of rest. Smolin believes a preferred state of rest exists that allows the definition of a preferred state of rest.

According to Smolin at each point of space there will be one special observer who sees the galaxies moving away from him at the same speed, possibly somewhere close to the vicinity of the original explosion, the big bang. Smolin also suggests that a preferred observer would see the cosmic background radiation with no variation in temperature at any direction. Connected with the preferred observer is a preferred time, a partial return to the universal time of Isaac Newton. Smolin claims general relativity can be recast as a theory with a preferred notion of time and synchronization; but that the synchronization depends on the distribution of matter and radiation throughout our universe.

Time Stops

I woke up and found myself floating around my room;
the legislature had given gravity a holiday.

If time is extended from geological and astronomical time
to cosmological time then it makes sense to talk about the far
future when all matter, black holes, etc. have dissolved into the
cold photons of the end of the universe. Roger Penrose points
out this condition is virtually the same as the supposed
beginning of the universe except the collective temperature of
the end universe photons are far lower that the collective
temperature of beginning universe photons, assuming it
makes any sense at all to talk of an emergent quantity such as
temperature in these kind of conditions. Penrose also
conjectures that time stops at the end of our universe since it
is impossible to construct a clock out of mass less photons.
After all, how can there be time if there are no possible clocks?

But just because you can't measure something doesn't
mean it doesn't exist. Evidently to both Penrose and Bohr, if
you can't measure something such as time or the definite
position of a particular particle, it makes no sense to talk
about time or the definite position of a particle. This is a
philosophical point basically. What does measurement have to
do with existence; especially since measurements are

something defined within a particular theoretical framework, quantum theory in the case of Bohr.

If something can't be measured, then it has no place in science, and if something has no place in science then it makes no sense to talk about it at all. One of the strongest arguments against string theory is that it has made no new predictions that can be measured; in other words no one knows if it is correct or not. But on the other hand, perhaps string theory is correct.

The point is to admit ignorance rather than throw away all possibilities. In the case of the end of the universe it's hard to say whether time as we know it exists at all. Perhaps it exists, perhaps it doesn't.

Time's Arrow

The rumor is the inventor of the Panopticon had himself stuffed after death so he could continue to attend committee meetings.

A gas in equilibrium inside a box fluctuates around some particular temperature and pressure, emergent quantities which describe the gas as a whole. Temperature and pressure do not describe the particular motions of the atoms and molecules that comprise the gas, they are averages that apply to the entire gas. On the other hand, that doesn't mean that temperature and pressure don't exist and that it means nothing to talk about temperature and pressure. In the gas of a system in equilibrium such as the gas box it is hard to distinguish one time from another since the fluctuations in the gas's states are random; however, physical systems that are not in equilibrium often flow in one direction in time. The universe is not in equilibrium because it is continually expanding, which gives a cosmological arrow of time. The entropy of most isolated systems increases, which is called the thermodynamic arrow of time. Living things are born, grow, and die, they do not die, grow, and then are born. This is called the biological arrow of time. We always experience time flowing into the future, never into the past; another form of the arrow of time. Light waves always move into the future, which means there is an electromagnetic arrow of time.

Lee Smolin takes the existence of an arrow of time to imply the fundamental reality of time; although the precise logic behind statements of this kind seem lost in a kind of philosophical fog familiar to anyone who has attempted to read or make sense of great philosophers. The philosophical fog seems to arise from a swamp of words and magical logical connections beyond the ken of ordinary mortals who don't claim any ability to perceive absolute truths. Mr. Smolin has accepted a position as an adjunct professor of philosophy at the University of Toronto in addition to his duties as one of the founders of the Perimeter Institute at the University of Waterloo, which works on loop quantum gravity in addition to other programs at the edge of modern physics. Whether this edge of physics is a leading or receding edge is a matter of debate; however, Smolin's interest in philosophy is unusual in that most mathematical physicists would never admit to an intellectual pursuit such as philosophy that appears to be analogous to an undue interest in freak shows at carnivals.

Emergent Time

I ordered a double monster chili reality burger at McBullshits; it tasted like cardboard.

It makes no sense to talk about the temperature or pressure associated with a single particle moving through space and time; similarly it makes no sense to talk about the space and time associated with a single particle unless the motion of the particle is related to an outside frame of reference. But the use of an outside frame of reference implies an outside observer, and we slide into the same problems with solipsism currently dogging the interpretation of measurements in quantum theory. However it does make sense to talk about the temperature and pressure of a group of gas molecules in a room, just as it makes sense to talk about the motion of a single particle against the background of a thermodynamic system or of the entire universe.

The isolation of a single part of a system can result in the loss of information. Although we tend to think of elementary particles as "fundamental" and try to construct entire systems from their properties, we slough over the exact manner in which emergent information such as temperature, pressure, and time are related or emerge from the single parts. The path from quantum theory to classical relativity theory is filled with hand waving, conjecture, and exceedingly rare admissions of ignorance.

It may be more important to understand how time emerges from a physical system that to understand whether or not time is fundamental or "real".

Math and Physics

South Korea must be the biggest country in the world; the U.S. arrived in 1950 and still hasn't found its way out.

Many of the statements in Lee Smolin's "Time Reborn" would strike a mathematician as wild and wooly; but theoretical physicists have never valued precision and clarity as much as mathematicians. However, according to Peter Woit in his book on the failure of string theory "Not Even Wrong", the two biggest sources of inspiration for mathematicians are numbers and mathematical problems presented by physical theory just as mathematics provides a method of tackling physical problems. The most prominent case of the use of mathematics in physical problems from the twentieth century is probably the use of tensor calculus in general relativity or the use of differential eigenvalue theory in quantum theory. Peter Woit suggests string theorists should adopt the attitude of mathematicians at the very least since their discipline has no physical reality as a check; and that most string theorists should simply give up.

A common puzzle is why mathematics should describe reality at all. Although we have some idea of the essence of mathematics since it is a human creation, we don't necessarily have any idea of the essence of reality even though some upper class elites seem to think they create reality at their whim. Perhaps a clue to this puzzle is the misuse of mathematics in modern economics. In economics mathematics is sometimes used to justify government policies or the policies of banking elites; mathematics can be misused just like any other tool. The fact that mathematics is such a useful tool in physical sciences is due to the brilliant but careful manner in which it is used; however mathematics is not a universal tool or panacea, good for any ailment.

Peter Woit wonders if a universal physical theory could be constructed by studying the representation theory of gauge symmetries and the diffeomorphisms in the tensor calculus of general relativity. This is a classical approach melding mathematics and physics somewhat reminiscent of Newton's virtually simultaneous achievements in calculus and astronomical mechanics. On the other hand, perhaps quantum theory itself should be recast in a form that includes quantum gravity including a more comprehensive understanding of the way time emerges from various physical systems.

The real problem does not seem to be any mismatch between mathematics and physics; but between physics and philosophy. Most physicists believe philosophy has nothing to do with their subject and has nothing to say about their problems. As examples, quantum theorists swallow the solipsism implicit in the Copenhagen interpretation of their theory without hesitation while cosmologists make grand leaps of reasoning by postulating dark matter and energy without invoking any philosophical school at all. The assumption seems to be that their reasoning is rational, whatever that means.

Failure

Barbed wire and armed guards surround every free speech zone. Once inside the prisoner may never get out or even know the charges.

While discussing the failure of string theory Daniel Friedan said: "Recognizing failure is a useful part of the scientific strategy. Only when failure is recognized can dead ends be abandoned and useable pieces of failed programs be recycled. Aside from possible utility, there is a responsibility to recognize failure. Recognizing failure is an essential part of the scientific ethos. Complete scientific failure must be recognized eventually."

Friedan's view on failure should be extended to quantum theory. Quantum theory is a philosophical failure and should be recognized as a failure. As mentioned before, an entire journal devoted to rectifying this failure exists: "Foundations of Quantum Theory". Unfortunately contributors to this journal seem to be people outside the mainstream of theoretical physics.

In addition it should be admitted that relativity theory fails to predict the various movements of galaxies. The postulations of dark matter and energy are attempts to keep relativity theory in its current form without making any major changes in the role of synchronizing radiation, i.e. light or in the role of time.

Oil Droplets

Heretics who refuse to accept Copenhagen are lined up against the blackboard and machine-gunned. In their next life they are cab drivers and cooks.

In 1932 John von Neumann, a later favorite of the American military-industrial complex, supposedly claimed to have proved that the probabilistic wave equation of quantum theory could have no hidden variables; nearly 30 years later John Stewart Bell proved this wrong. In addition, Yves Couder at the Paris Diderot University conducted experiments in which an oil droplet bouncing on the surface of a silicon oil fluid showed properties of a quantum mechanical particle.

According to Natalie Wolchover's "Have We Been Interpreting Quantum Mechanics Wrong This Whole Time?" in the 6/30/2014 issue of Quanta Magazine:

"The experiments involve an oil droplet that bounces along the surface of a liquid. The droplet gently sloshes the liquid with every bounce. At the same time, ripples from past bounces affect its course. The droplets interaction with its own ripples, which form what's known as a pilot wave, causes it to exhibit behaviors previously thought to be peculiar to elementary particles-including behaviors seen as evidence that these particles are spread through space like waves without any specific location until they are measured."

In pilot wave theory, first suggested by de Broglie, and developed further by David Bohm, particles would set up fluid like undulations in space-time and would interact with those undulations. If the space-time fluid were a fluid without dissipations, a superfluid, entanglements between two particles could be explained by assuming the initial interaction between the two particles permanently affects the contours of the space-time superfluid. Also, it may be that pilot wave theory, or deBroglie-Bohm theory, is the only possible entry we have to a valid quantum gravity.

Eigenvalues

One of the favorite oxymorons of capitalism is 'sustainable development'.

Although the oil droplet experiments seem to lend credence or plausibility to pilot wave interpretations of quantum theory, the droplets are very large physical objects with respect to atoms and elementary particles. However, there are other arguments in favor of pilot wave quantum theory.

In John Von Neumann's 1932 "Mathematical Foundations of Quantum Mechanics" the mechanics of quantum theory is formalized as an axiomatic system in which quantum mechanics was reduced to the mathematics of linear Hermitian operators on Hilbert spaces. There is some question whether this formulation serves as an obfuscation rather than a clarification. Von Neumann's work on quantum mechanics, which he considered one of his most important contributions, depends on the use of eigenvalues, eigenfunctions, and Fourier series in solutions to partial differential equations. One of the primary physical assumptions is that the result of any measurement, whatever a measurement is, is an eigenvalue that appears as one of the numbers that appears in an expansion of the wave function, the solution to the wave equation. For this to work physically, at least in the Copenhagen interpretation, the wave equation must collapse on measurement to an eigenfunction corresponding to the particular observable being measured. As a matter of fact, wave function collapse is necessary for a measurement, whatever a measurement is. According to Von Neumann, the entire universe can be described by a universal wave function describing everything. Any measurement pertaining to the universal wave function must cause it to collapse. Von Neumann assumed the collapse in this case, and therefore in every other case, is caused by the consciousness of the observer. Although Eugene Wigner accepted this logic, many physicists including John Bell did not accept it. Bell, in discussing the existence in Von Neumann's formalism of a 'movable boundary' between the classical measuring apparatus and the measured quantum system, said:

"A possibility is that we find exactly where the boundary lies. More plausible to me is that we will find there is no boundary. The wave function would prove to be a provisional or incomplete description of the quantum-mechanical part, of which an objective account would be possible. It is this possibility of a homogenous account of the world, which is for me the chief motivation of the so-called 'hidden variable' possibility."

Ancient Disputes

The world does not exist because an observer sees it; otherwise, why would all of us see the same thing, more or less.

Grete Hermann discovered the flaw in Von Neumann's great 1932 work in quantum mechanics in 1935; many years later Bell rediscovered the same flaw with respect to the exclusion of hidden variable theories. However, in 2010 Jeffery Bub maintained Von Neumann's proof doesn't attempt to prove the absolute impossibility of hidden variables and therefore is not flawed; however, there is no record of Von Neumann attempting to correct the near universal misinterpretation for over 30 years that hidden variable theories are mathematically impossible.

Bell, on the other hand, was not so much interested in the mathematical precision offered by an axiomatic approach as in physical precision. According to Bell:

"For the good books known to me are not much concerned with physical precision. This is clear already from their vocabulary. Here are some words which, however legitimate and necessary in application, have no place in a formulation with any pretension in physical precision: system, apparatus, environment, microscopic, macroscopic, reversible, irreversible, observable, information, measurement. On this list of bad words from good books, the worst of all is 'measurement'".

For example, it's not exactly clear what the difference between microscopic and macroscopic is. To say that microscopic is defined by that region small enough so that quantum mechanics applies, and macroscopic is that region large enough so that classical mechanics applies is not very helpful. The question arises how one passes from microscopic to macroscopic, and how to explain large scale 'quantum effects' such as superfluidity, superconductivity, and black holes. Even worse, from a philosophical point of view, are the words 'observable' and measurement'. Bell was interested in a philosophically robust, "observer free" quantum mechanics. According to Bell, physical theories should not be concerned with observables, but be-ables. From Bell again:

"The be-ables of the theory are those elements which might correspond to elements of reality, to things which exist. Their existence does not depend on 'observation'".

E.P.R.

Although Einstein made major contributions to theoretical physics he believed quantum mechanics was incomplete, a kind of senile delinquent thought.

In 1935, Einstein, Podolsky, and Rosen proposed their famous thought experiment in which a single system splits into two that travel in different directions. They reasoned a measurement of one of the two parts would be communicated instantly to the other part; a violation of relativity that argues no physical cause can propagate faster than the speed of light.

Einstein had the same philosophical view as Bell and protested against the idea that there exists no objective physical reality other than that which is revealed through measurements interpreted by the quantum mechanical formalism of Von Neumann. After Bell's work with the hidden variables problem, and the development of his well-known inequality, Alain Aspect of the Institut d'Optique d'Orsay was able to give flesh to the EPR thought experiment by considering radiative cascades in unstable atoms. The radiative cascades resulted in photons that were entangled with each other in the same way as the hypothetical two parts in the EPR system. The results confirmed standard quantum mechanics. According to Aspect:

"Entanglement is definitely a feature going beyond any spacetime description a' la' Einstein: a pair of entangled photons must be considered a single global object, that we cannot considered as made of individual objects separated in spacetime with well-defined properties."

Unfortunately, this in turn brings up a nasty cosmological problem. If everything was created in the big Bang, isn't everything in the universe entangled with everything else? How is it possible to tell one physical object from another?

Renormalization

To agree with experiments, we have to dress particles properly.

Sean Carroll of Caltech reports some interesting statistics in his blog preposterousuniverse.com from a graph Carroll calls "The Most Embarrassing Graph In Modern Physics" that was taken from a poll by Schlosshauer, Kofler, and Zeilinger. The graph indicates 42% of polled physicists agree with the Copenhagen interpretation of quantum theory, 18% agree with Everett's many worlds interpretation, and 40% agree with other interpretations such as Von Neumann's and Wigner's. He's right, it is extremely embarrassing because it means after nearly ninety years there is no consensus on the interpretation or philosophical meaning of quantum theory. However, the usual excuse for this situation is that quantum mechanics predicts so much so accurately. This is not a good argument for any lack of understanding; and there are several situations where the application of quantum mechanics is tortured and manipulated to produce correct results such as in decoherence theory and renormalization calculations. In renormalization theory results agreeing with experiment, the Lamb shift for example, require a second quantitization to include the effect of virtual charges on the measured charge of a particle such as an electron. The experimental electron is a "dressed" electron. However, as mentioned before in this essay, many particle descriptions such as second quantitization for just the single electron is a global description, and it cannot be a global description because spacetime is a fixed Minkowski space in this description, fixed by fiat or initial assumption.

Global

Globocop is now Robocop too with its drones.

The violations of Bell's inequality rule out many local hidden variable theories or explanations of quantum mechanics; but these violations do not rule out or prohibit global hidden variable theories such as deBroglie's pilot wave. The pilot wave is a global variation of the space in which a small quantum particle travels.

In addition, entanglement is global and involves at least two particles. But does the big Bang theory really imply that everything is entangled with everything else since, hypothetically, everything was created together at the beginning of time for our universe out of nothing by a huge quantum event? No. Not necessarily.

There is no comprehensive theory of quantum entanglement just as there is no theory of quantum gravity. As far as we know virtually anything is possible. Perhaps part of the matter and radiation created during the big Bang behaves like current matter and radiation while there are many other kinds of matter and radiation created; and perhaps the different kinds of matter and radiation were not even created at the same event or at what we could recognize as the same time. In addition, perhaps different types of matter and radiation superimpose on one another in the macroscopic world we are capable of observing, and give us quantum tunneling, interference patterns, stable atoms, super-fluids, and everything else we see or think we see. If there were a comprehensive theory of entanglement perhaps it would be clear what happened at the beginning of time.

It appears the reason there is no comprehensive theory of either entanglement or quantum gravity is that quantum mechanics, as it now stands, cannot be extended. What is the reason for this situation?

Matter and Radiation

Are you now, or have you ever been, a serial thought criminal?

In 1905 Einstein published a paper called "Does the Inertia of a Body Depend on its Energy -Content" in the Annals of Physics. He used a conservation of energy calculation to describe a particle or body before and after emitting radiation in the form of light. Light carries energy and the amount of energy carried varies depending on the motion of the system in which it is measured. Einstein considered conservation of energy in both a coordinate system at rest with respect to the particle and in a moving system. The result was his famous and very short formula.

It makes no sense to question conservation of energy, but the energy of any emitted radiation could easily not transform like light. Even if the emitted radiation were light, light might not transform in the same manner as light in our immediate vicinity or at our scale. Light might not even travel at the same speed at extremely small scales or in some of the more extreme parts of our universe. The Michelson–Morley interference experiment is given wider application than warranted. What kinds of applications are warranted by any physical experiment, for that matter? Could Einstein's calculation be done in a somewhat more general fashion for radiations in general or for a particular class of radiations transforming in a more general fashion, a more universal kind of "light"?

Foundational Questions Institute

The dyslexic paranoid was convinced he was following someone.

The Foundational Questions Institute is asking for essays on the relationship between physics and mathematics this year:
"In many ways, physics has developed hand-in-hand with mathematics. It seems almost impossible to imagine physics without a mathematical framework; at the same time, questions in physics have inspired so many discoveries in mathematics. But does physics simply wear mathematics like a costume, or is math a fundamental part of physical reality?

Why does mathematics seem so "unreasonably" effective in fundamental physics, especially compared to math's impact in other scientific disciplines? Or does it? How deeply does mathematics inform physics, and physics mathematics? What are the tensions between them—the subtleties, ambiguities, hidden assumptions, or even contradictions and paradoxes at the intersection of formal mathematics and the physics of the real world?

This essay contest will probe the mysterious relationship between physics and mathematics."

Mathematics in some of the social sciences such as economics and sociology doesn't even appear as a costume, but as a nasty means of obfuscation, a way to give unwarranted credence to academic bullshit, another way to grind an ideological axe. And even in the case of Von Neumann's "Mathematical Foundations of Quantum Mechanics", the use of eigenvalue theory appears to lead Von Neumann to believe in the power of observation to define physical reality. Von Neumann's masterpiece was published in 1932, and it pretty well nailed down the mathematics of quantum mechanics; and its prestige may sink the possibility of extending quantum mechanics to cover quantum gravity.

Mathematics and Logic

According to the Red Queen hypothesis (running harder to stay in one place), coevolution of hosts and parasites selects for sexual reproduction of both.

Early in the twentieth century, Bertrand Russell attempted to tame mathematics by describing a set of axioms and inference rules in symbolic logic from which all mathematics could be derived; thereby bringing mathematics entirely within the systems of thought commonly called philosophies. His great work was called Principia Mathematica, or PM, for short. From Wikipedia:

"PM, as it is often abbreviated, was an attempt to describe a set of axioms and inference rules in symbolic logic from which all mathematical truths could in principle be proven. As such, this ambitious project is of great importance in the history of mathematics and philosophy, being one of the foremost products of the belief that such an undertaking may be achievable. However, in 1931, Godel's incompleteness theorem proved definitely that PM, and in fact any other attempt, could never achieve this loftly goal; that is, for any set of axioms and inference rules proposed to encapsulate mathematics, either they system must be inconsistent, or there must in fact be some truths of mathematics which could not be deduced from them."

If logic and mathematics don't fit together all that well by themselves, how can the two be up to a description of physical reality? First of all, what is physical reality, and why should we expect mathematics and logic to describe whatever physical reality is?

Perhaps the only cogent assumption is that physical reality exists aside and outside of ourselves, which precludes the opinions of Wigner and Von Neumann.

Axiomatic Systems

"I always thought Von Neumann's brain indicated he was from another species, an evolution beyond man."

Hans Bethe

The most famous axiomatic system from antiquity is Euclid's geometry, which appears to have had such great influence in mathematics if not philosophy that everyone, even into current times, has tried to emulate it. Euclid's geometry involved idealized, mathematical shapes and forms. Von Neumann's axiomatic system, the "Mathematical Foundations of Quantum Mechanics", on the other hand involves mathematical rules of thumb used to explain very odd, mysterious, and even inexplicable physical behavior, from our relatively large point of view. Should an axiomatic system be used to describe a physical reality that has no widely accepted interpretation among scientists? Doesn't an axiomatic system attempt to set up rigid rules and make laws out of simple rules of thumb, rules that may work in some circumstances but not in others?

There are holes in quantum mechanics: its interpretation is a mystery, it does not mesh with relativity, there is no working theory of quantum gravity, and there is no smooth path from microphysics to macrophysics since quantum statistics doesn't apply in every case such as superconductivity.

There are no such holes in relativity theory except, perhaps, for the specialization to the particular form of light in our region; however in some of the attempts to reconcile quantum theory and general relativity, changes are proposed in general relativity rather quantum mechanics. What kind of sense does this make?

The End

"There is a real world, and it is essentially the same for all of us."

"Smart is better than dumb."

Verle's two principles.

This is the end of many years of speculation, conjecture, and total bafflement. I give up.

One of the rules of thumb that appear to be useful outside of any axiomatic structure is the idea of superposition; in addition, extending relativity to include general classes of radiation or "light" may be useful. Perhaps a complete version of quantum mechanics would be based on a superposition of various forms of light, a multifoliate-time quantum mechanics. At any rate, trying to extend the current incomplete version of quantum mechanics without cleaning up its philosophical and technical bases appears to be a mistake leading to nothing but dead-ends and abject failures.

.

www.ingramcontent.com/pod-product-compliance
Lightning Source LLC
Chambersburg PA
CBHW080819170526
45158CB00009B/2470